Methuen Drama Modern Classics

Methuen Drama's *Modern Classics* series showcases landmark plays from around the world. Drawing on the *Modern Plays* series, which launched in 1959, *Modern Classics* celebrates plays from the contemporary repertoire by world-leading dramatists and presents their work in a definitive edition, alongside new introductions by leading scholars and industry professionals. With writers such as Pulitzer Prize-winners Jackie Sibblies Drury, Ayad Akhtar and David Mamet through to Lucy Prebble, Katori Hall and Caryl Churchill, *Modern Classics* are ideal for students and anyone wanting to deepen their knowledge of the plays that form part of the modern dramatic canon.

Deep Blue Sound

Rich, funny, and devastating. As Abe Koogler's sweet, sad village mosaic revealed itself heart fissure by heart fissure, silence spoke volumes. – Observer

A small island community in the Pacific Northwest comes together to search for answers when a beloved pod of whales goes missing. As the seasons pass and the search grows cold, the islanders struggle to work together, while grappling with changes in their own lives.

Abe Koogler's play is a warm and surprising portrait of a group of idiosyncratic people looking for connection and meaning in the face of loss.

After an acclaimed remount at New York's Public Theater, *Deep Blue Sound* is published by Methuen Drama's Modern Classics series, featuring a new introduction by Crystal Finn.

Abe Koogler's plays have premiered at Playwrights Horizons, Manhattan Theatre Club, and Lincoln Center Theatre. He won an Obie Award for Playwriting, the Dramatist Guild's Lanford Wilson Award, and Williamstown Theatre Festival's Weissberger Award, among other honors. He earned graduate degrees in playwriting from UT-Austin and Juilliard. He teaches playwriting at Bennington College and also works as a political speechwriter.

Deep Blue Sound

ABE KOOGLER

With an introduction by
Crystal Finn

methuen | drama
LONDON • NEW YORK • OXFORD • NEW DELHI • SYDNEY

METHUEN DRAMA
Bloomsbury Publishing Plc, 50 Bedford Square, London, WC1B 3DP, UK
Bloomsbury Publishing Inc, 1359 Broadway, New York, NY 10018, USA
Bloomsbury Publishing Ireland, 29 Earlsfort Terrace, Dublin 2, D02 AY28, Ireland

BLOOMSBURY, METHUEN DRAMA and the Methuen Drama logo are trademarks of Bloomsbury Publishing Plc

First published in Great Britain 2024
This edition published in the Modern Classics series 2026

Cover design: Ben Anslow

A catalogue record for this book is available from the British Library.

A catalog record for this book is available from the Library of Congress.

ISBN: PB: 978-1-3506-0102-4
ePDF: 978-1-3506-0103-1
eBook: 978-1-3506-0104-8

Series: Modern Classics

Typeset by Newgen KnowledgeWorks Pvt. Ltd., Chennai, India
Printed and bound in Great Britain

For product safety related questions contact productsafety@bloomsbury.com.

To find out more about our authors and books visit www.bloomsbury.com and sign up for our newsletters.

Contents

Introduction, by Crystal Finn vii
Deep Blue Sound, by Abe Koogler 1

Introduction

In 2023, and then again in 2024, I was lucky enough to be cast in *Deep Blue Sound*, Abe Koogler's beautiful play, which was produced by Clubbed Thumb, and is set in and amongst an island community off the Pacific Northwest.

Participating in both productions was a thrilling and profoundly gratifying experience for me.

It was also a hugely *embarrassing* one, for a singular reason, which was: the number of times during performances (for paying audiences!) when I broke character and openly laughed on stage like a complete amateur.

I'm not talking about the kind of laughter that comes when someone in the audience answers their phone and talks to their therapist for a couple of beats, or when someone farts on stage—I'm talking about unprompted guffawing from simply saying one of *my own* lines on stage. This kind of corps-ing is not only totally unprofessional, it is also hard to justify in a play which is about, among other things, death, ecological collapse, and endemic loneliness.

In my defense: Abe is the funniest writer I know, and *Deep Blue Sound* is, arguably, his funniest play, and just because it *happens* to be about death, ecological collapse, and endemic loneliness, doesn't mean it is not also hilarious, because I don't think Abe could write something unfunny if he tried.

And also, *maybe,* in my defense, the play isn't funny in spite of those themes, but because of them, since Abe's humor, like any humor that is worth anything at all, is the lashed-to-the-mast-of-humanity variety, the variety that makes you laugh and *then* makes you cry, because it is so undeniably true.

Actually, crying was a problem too. Crying on stage when you shouldn't be crying! I did it. My castmates did it. We got yelled at. We were told: *you can't cry here, your <u>character</u> isn't crying … But that line!* We would say, *it makes me think of the time when … It makes me remember my <u>own</u> mother when she …*. And the director would stop us and remind us we were doing a play.

So it wasn't our fault! We were confused. The play felt too real.

The characters in *Deep Blue Sound*, all of them members of the same small community, all of them trying their best to be good members of that community, feel so real because they are surprising. They don't follow plots, or themes, or revelations. They follow their own idiosyncratic needs, desires, prejudices, love. These are full people who act on their own impulses, willfully and heartily, no matter the time allotted them in a scene, no matter whether they are major players or barely mentioned by anyone.

I'm thinking of one character who shows up at a Christmas sing-along who *really* wants to have a beer with another character. That's all we know about her! She doesn't have a name, or a plot point. She exists in a glimpse of peculiar intimacy—peculiar to her—which is to say, real to her. She's not a character from a *play*; she's a character from life—from some potluck your parents dragged you to when you were eleven, some woman you sat next to on a plane one time.

Or take the town therapist: we know nothing about her life. She is just a voice offstage. But she is such a *good* therapist. She says the right things—the things an actual therapist would say. You can tell she takes her job seriously. She is warm, but firm. She reminds me a little of a teacher I once had who—oh no, I'm just projecting again. From my own life.

That's what the play does to you. It cracks open the fourth wall—invites all the people you know to join you on stage and suddenly you're laughing and crying when you shouldn't be because you're thinking of your own mom, your own Grandma, your own acting teacher from childhood.

And then …

Well … the *play* does something funny with these very real characters.

Something almost unbearable.

It picks these characters up from their grounded, real lives, and it drops them into a strange, dark ocean, and it tells them to swim.

All great plays move the form of theater forward in some way. They teach us something new about how to watch plays, how to *act* in them. What *Deep Blue Sound* does, that no other play has done before it, is it gives us characters utterly true to life, of the earth, quotidian, small even in their ordinariness, and places those characters in a *composition* that is the very opposite of those things, that is in fact, cosmic.

That composition, which you will experience on the page when you read the play, is a swift-moving, hang-gliding swoosh of language, which doesn't just gesture toward themes of temporality, but also, or maybe rather—mimics the experience of that tempo itself. Scenes move quickly in this play, which is not to say they should be played fast—but rather that they are always and precisely left unfinished. Two characters may be just on the verge of figuring something out about each other, when we are pulled away, plunged forward, never learning what is being truly said, truly meant. The end of one scene is always the beginning of another—so that interruption takes on the feeling of continuity, the result of which is a fluidity of rhythm between disparate stories, disparate needs. We switch topics with no warning, we switch seasons with no warning, we switch identities with no warning. All these small, beautiful moments, existing in an expanse of blue which the play insists, lovingly, but firmly, is ferrying us along at an untraceable pace—toward an unknown end.

The play is so so small in its familiarity and intimacy, and so so expansive in its engagement with time. In this way also it is exactly, indelibly, like life.

Other plays have given us similar magic tricks. I'm thinking of *Our Town* primarily, which many people mentioned to me after seeing *Deep Blue Sound.* That play also centers on a small town; it also is fundamentally about time and how to be aware of it.

Do any human beings ever realize life while they live it? Emily asks.

No is the answer.

The question is one of proportions really: how do I live? In the big or the small? In *Our Town* Wilder gives us that question, by way of a guiding stage manager, who takes us over the threshold of the living, and into the land of the dead.

In *Deep Blue Sound* there is no guide. The question is asked in the very structure of the play itself.

It is asked by placing two different metronomes in one play, and letting them play against each other in a reckless kind of battle for dominance.

Which will win out: the day-to-day exchange between one man and another sitting on a porch talking about some day from the 90's? Or the merciless and unbreakable wave of infinite epochs?

In the end, neither, since there is no way to separate out lived-daily life from the grander meaning it carries, just as in Abe's play there is no way to separate out the importance of small, individual exchanges, from the tapestry they are a part of.

Three pages from the end of the play, however, for one character, a woman named Ella, who up until this point in the play has been denying her cancer diagnosis to her friends, and to herself, the two tempos of the play do get very close together. For Ella, in fact, in the last moments of her illness and life, I think maybe they move close enough that they become, for a brief second, truly and transcendentally indistinguishable.

She is in her room, having chosen the day which she will die, helped by the aid of a nurse, and her best friend John has just delivered flowers to her. When he leaves she wakes up for the last time. She looks around her. She says:

> I'm awake
> I can see the flowers
> I can see outside the room
> The house ... the garden ... the road
> Luckily I can see through my eyelids now
> Luckily I'm hearing through my fingers now
> Luckily my head is opening towards the sky

She is dying. But she is not dead. Her inner eyes are open. They see something the rest of us don't: they see the flowers, then the room, the house, the garden, up the road … and on and on past the horizon … to the sky. Here finally the small becomes large, the fast becomes slow, the cosmic becomes … ordinary.

In our production Ella was played by the incredible actress Maryann Plunkett—maybe the most grounded and human of all human actresses alive, and listening to her experience this moment each night was hard. All of us in the cast were onstage by then, surrounding Maryann in different seats, and we all had various strategies for not looking like we were crying when she said those lines: hair over the face, hand over the mouth.

What were we doing there? Acting? Not really, no. *Trying* to *keep* acting so that we wouldn't be overwhelmed by living? We were neither acting, nor living, but just watching—which is the privilege of any audience—to leave the time signature of your own life for a few hours, to remember it all the better when you return to it.

Sometimes to keep my emotions in check during her speech I found myself swooping my gaze over the actual audience, and on some nights I would swear I saw someone out there in the house that I knew—not a friend from the *theater*, or an acquaintance from the neighborhood—*no*, I mean someone from my childhood, someone from my *hometown*: my mom, my Grandma… my first acting teacher. And then, as quickly as they appeared, they would be gone, pulled into the dark, into the past, swept along into the vast ocean, where I like to think so many of Abe's characters are still swimming.

If you get a chance to act in this play, my advice is this: don't cry, don't laugh. But if you do, remember it's only because you are alive and living and swimming in very deep waters. Enjoy it.

Deep Blue Sound

Deep Blue Sound opened at Clubbed Thumb in New York City (Maria Striar, Artistic Director; Michael Bulger, Producing Director), where it had its first performance on June 5, 2023.

Ella	Maryann Plunkett
Joy Mead	Natsuko Ohama
John	Thomas Jay Ryan
Mary	Tala Ashe
Chris	Armando Riesco
Ali	Brittany K. Allen
Mayor Annie	Crystal Finn
Leslie	Jan Leslie Harding
Gary	Bruce McKenzie
Director	Arin Arbus
Set	dots
Lights	Isabella Byrd
Costumes	Emily Rebholz
Sound	Mikaal Sulaiman
Choreography	David Neumann
Props	Natalie Carney
Stage Managers	Caroline Englander, Matthew Pezzulich

Characters

Ella, *sixties, dying of cancer (this actor also plays* **Islander F***)*

Joy Mead, *fifties to seventies, the newspaper editor (this actor also plays* **Islander D***)*

John, *fifties to sixties, used to run an art gallery on the island (this actor also plays* **Islander G***)*

Mary, *thirties to forties, a massage therapist (this actor also plays* **Islander A***)*

Chris, *thirties to forties, a carpenter (this actor also plays* **Islander E**, **Alexander the Dancing Boy**, *and* **Voice of an Oceanographer***)*

Ali, *twenties to thirties, daughter of Ella (this actor also plays* **Islander B** *and* **Karl Who Runs the Pharmacy***)*

Mayor Annie, *the unofficial mayor (this actor also plays* **Islander H** *and* **Mo**, *the mother of Alexander)*

Leslie, *a horse groomer (this actor also plays* **Islander C** and **Voice of a Therapist***)*

Gary, *a drifter (this actor also plays the* **Whale***)*

Don't worry too much about keeping track of all this doubling. Whether you're reading the play or performing it, just dive in! You'll figure it out as you go.

Actors can step in and out of roles without too much rigamarole. In general, no costume changes are needed. Things may not be as clearly delineated as we're used to.

** indicates a change in location or time, or just a shift to another character. Mostly these shifts should be instantaneous, as if the play is one long connected thought.*

Time

The present.

Location

An island of several thousand people in the Pacific Northwest, accessible only by ferry boat.

Scene One

A moment when everyone is still.

Then:

Islander B The whales!

Islander A We're gonna tell you what's going on with the whales.

Islander C The whales? I thought . . .

Islander G What's going on with us. (And the whales.)

Islander H What's been happening here on this island.

Islander C In this moment.

Islander F In our lives.

Islander C (*with great seriousness*) In *this* moment . . . in our lives.

Islander B They just said that.

Islander H About the whales *and* about sort of *us*. Islanders.

Islander G Everyone plays more than one person.

Islander F I play Ella, who's dying of cancer.

Islander G I play John, her dear friend who doesn't know that.

Islander A I play Mary, the three of us are friends, and I just split up with my husband Chris.

Islander E Who's not doing so good right now. I play Chris.

Islander C I play Les, or Leslie for long: a horse groomer with a new pen pal.

Islander H I play Mayor Annie. *Recently* elected.

Islander B I play Ali. Just moved back from Brooklyn to take care of my mom.

Islander D I'm doing Joy Mead. Editor of the paper. Just takin' it all in.

Islander B Secondary roles! I'm *also* doing Karl who runs the pharmacy, a local business.

Islander H I play Mo, a local mom. That's all we know about Mo.

Islander E I'm playing Alexander her son who *just* wants to be a dancer.

Islander C (*into a microphone so we just hear her voice)* I play a therapist with a practice on the island who *tries* to be discreet when I see my clients out and about.

Islander E (*into a microphone so we just hear his voice*) And I play a German oceanographer currently living in Switzerland for work. Many years ago I had intercourse with someone in this group.

Islander B That's who we're playing. And we're all a separate third thing.

Islander F Islanders.

Islander C People who live here. The people you saw walk in.

Islander A A lot to keep track of. Don't try and keep track of it. Just give up now.

Islander G If you get mixed up, let it go.

Islander F What you need to know, you'll come to know.

Islander H One thing you *do* need to know: we are *all* worried about those whales.

Islander E Where the HELL are those whales?

Islander B Okay let's do a general overview of us, circle back to the whales. Okay?

Islander E . . . Okay.

Islander D Sure.

Islander G Yup.

Islander H Yah.

Islander A We've been living here, on this island

Islander F Some of us longer than others **Islander G** Some of us for decades now

Islander C Some of us were descended from those who'd settled it

Joy Mead And some of us had moved over from the city after our divorce **Islander E** For the water and the woods . . .

Ali And some of us had come back because our mom got sick

Islander H But you were here as a kid

Ali Yeah but I left.

Islander H You left and had a great career

Ali Well

Islander H She went to New York, and had a great career, as an / artist

Ali Graphic designer

Islander H A brilliant career as a

Ali Freelance graphic designer

Islander H In New York!

Islander F The island is

Islander G We're talking Pacific Northwest, take a ferry boat to get here

Islander F Not too far from Seattle, can see the lights from the

Islander A Far enough to be somewhere really different

Islander E Thirty minutes or so by ferry—does something to ya

Islander D The water the light

Islander H In terms of layout there's a town center

Islander C We *have* a grocery store, and a bookstore, and I love to read

Islander E A couple *great* restaurants

Islander F I wouldn't say "great"

Islander G There's an art gallery . . . **Islander C** Historical fiction . . . you name it!

Islander D Where *I've* shown my work.

Islander H She does these wonderful, how would you . . .

Islander E They're conceptual . . .

Islander H They're very *dense*, they sort of . . . *collect*

Islander F But it's a beautiful island

Islander D (How've you been? Haven't seen you at the uh)

Islander A (Yeah I've been kind of . . . .)

Islander D (Sure, sure . . .)

Islander A (But I'll stop by, I should stop by . . .)

Islander D (Stop by, yeah, not the same without you . . .)

Islanders A & D (Sorry)

Islander H The grocery store is a major meeting place

Islander B Accidental meeting place, sometimes

Islander G You have to keep your head down, if you want to get in and out

Islander F You have to really *move*, to escape a conversation

Islander D For example, I was in there **Islander G** (Are you comfortable?)

Islander H —and we ran into each other, and we had a little tête-à-tête

Islander D And it was / good **Islander F** (I'm good!)

Islander H It was good to catch up!

Islander B But sometimes you see someone and you think oh no: I'm going to browse the soups intently. I'm going to see what *kinds* of *soups* they *have*.

Leslie It's funny because people often do that to *me*

Islander B Oh yeah? Well, that's . . . .

Leslie And I'm Les, or Leslie for long, which is how I say it, and I'm a horse groomer! / (*Sudden anxiety:*) Did I say that already?

Mayor Annie And I'm the Mayor

Islanders A & B Annie is the "Mayor"

Mayor Annie I ran for it, and I won

Islander A It's not really a position

Mayor Annie It's an elected position

Islander A Well . . .

Mayor Annie People vote

Islander F Every year, at the Strawberry Festival

Islander D Long story short: it's a summer festival, there's a parade and booths.

Islander F The island is known for its wild strawberries

Islander G (Do you know I once worked for a summer as a migrant worker?)

Islander C (What a rich experience that must have / been!)

Mayor Annie The Strawberry Festival is where I was elected Mayor.

Islander A Symbolic mayor.

Islander C (*to* **Islander G**) You're always having *adventures*

Mayor Annie It's an *elected* position.

Islander A It's an elected symbolic position

Mayor Annie It's an elected position, with powers

Islander A There's no powers

Mayor Annie There *are* powers

Islander A Annie, but okay . . . what are the powers?

Mayor Annie To convene

Islander A Oh *okay*

Mayor Annie To bring people together, to start conversations. And I've done that

Islander A It's true you've started a lot of conversations

Leslie I always thought there was a point in my life at which I would go hiking . . . I mean throughout Spain.

Islander F Every year people vote a dollar based on which charity they want to support, and Annie was the candidate of the Island Humane Society

Mayor Annie I happen to love dogs

Islander F And so she won

Islander C And I ran on behalf of homeless youth, and I *didn't* win

Islander B Because there's no homeless youth on the island

Islander C Yes there was Rachel Braha's daughter Melissa who was um

Islander B She wasn't homeless **Islander F** Wonderful girl.

Islander E Pain in the ass

Islander C *Very* creative. Now *she* was creative.

Islander B She was having a fight with her mom, so she went to stay with Lisa Hartford

Islander E She went to school with my daughter—

Islander C She was *functionally* homeless

Islander E —and she was a pain in the ass

Islander C Well, *I* think there's a homeless problem, among youth, and that's just my opinion!

Islander B There's homeless adults, for example . . .

Gary Wood? Anyone need wood chopped?

Islander B No we're good

Islander E We're good, Gary, thank you

Islander G (That's / Homeless Gary)

Islander A Homeless Gary, he's

Islander D A little local color

Islander H Let's not condescend

Islander B He walks around the island carrying his chainsaw

Islander G It's a little threatening **Islander F** Have you ever hired him?

Islander A *He's* not threatening, he's just weird

Islander B He wouldn't use it **Islander H** That's condescending!

Islander A He just uses it to cut logs

Islander E That's true **Islander F** I would *hope* that's what he uses it for

Islander D I've hired him

Islander G Did he do a good job? **Islander C** Oh wonderful!

Islander D I'll tell you this: he was very thorough.

Mayor Annie The power to convene.

Islander G Oh right

Mayor Annie And I'm using it. For example to deal with the problem of stray dogs

Islander A There are a lot of stray dogs on the island **Islander B** *So many*

Islander F And no one knows where they're coming from **Islander H** There's not *that* many

Islander B (She just takes herself so seriously) **Islander H** A bit of background:

Islander E There are all these stray dogs on the island, that seem like wolves? Stray dogs that look wolf-y. They live in the woods.

Islander D My theory is that when Bob Hartman died

Islander C And that was a beautiful memorial

Islander F Thank you, I'd been practicing

Islander C Do a little bit

Islander F (*singing*) 'Tis a gift to be simple **Islander C** So beautiful

Islander D That when Bob Hartman died, his dog went out there and mated with the wolf.

Islander B There's a Wolf Haven on the island

Islander A There's a very strange woman named Star

Islander H Star's not that strange **Islander F** Star's wonderful, she's unique

Islander B A wolf sanctuary, she takes in strange wolves

Islander A Strange wolves?

Islander B *Stray* wolves, from around the country

Islander D Local color **Islander H** "Wolf Haven"

Chris She tried to fuck me once.

Islander G And a wolf escaped, and Star didn't want to admit it, for political reasons

Chris I mean fuck me *over*, I'm a carpenter? Didn't pay me.

Islander G She gets a lot of money from the government for those wolves

Islander F Which is a whole other

Islander G Turns out if you own a wolf you can make out scot-free on the taxes front

Islander F It's something to do with the Endangered Species Act

Islander H Oh wonderful!

Islander B So *I* think when Bob Hartman died, his dog went out there

Islander H The Endangered Species Act has been *so* important, *so* / important

Islander A It was sort of a gruesome

Islander G Sad **Islander E** Really sad

Islander B Case, of Bob Hartman

Islander A He'd been dead for a while, and sort of decomposing, and his dog

Islander F Don't tell this part **Islander H** Oh, come on

Islander A His dog had been eating? Um? Eating Bob Hartman?

Islander C Do we have to

Islander H (This isn't part of the main story, it's just background)

Islander D (What *is* the main story?)

Islander H (About the whales!)

Islander B And the dog, when they found it, was full-on crazy

Islander A I'm a volunteer paramedic, and it's true, the dog had gone crazy

Islander C And I'd met this dog **Islander G** *Sweet* dog, I mean originally.

Islander D Bob was a regular, around town, with this dog

Islander E Cup of coffee and a scone!

Islander F He was Mayor, for a year, actually **Islander B** (Good impression!)

Islander A (Symbolic mayor)

Islander F Running on behalf of veterans

Islander E He'd been in WW2

Islander B Cranky **Islander A** A complete asshole

Islander F He was *very* conservative

Islander E There's a difference between being conservative and being an asshole

Islander B Is there?

Islander E Hey now!

Islander H We respect your beliefs

Islander A *I* don't **Islander F** Do we? **Islander C** (I like that: "WW2")

Islander F The fact is that Bob was both conservative *and* an asshole. Two things that may or may not have been related.

Islander E But that's irrelevant **Islander B** They're related

Islander A So I get there, and the dog is eating Bob, or has recently eaten Bob because the dog has Bob *on his face*, and the dog snarls at me, and I thought *something has gone wrong with this dog, here in this closed house, with the windows shut*, and the dog ran away towards the woods, and *some* people think

Islander D That the dog mated with one of Star's wolves, who escaped years ago

Islander G She claimed it had died but everyone was like: where's the proof? SHOW US THE CORPSE

Islander B And now there's a problem, of little dogs that resemble wolves, and can be seen, sometimes, at night, in the woods, with their eyes.

Islander E I think they're just coyotes, come over from the mainland

Islander G How, they swam?

Islander E Animals are good swimmers

Islander C Well that's a generalization

Islander A Those aren't coyotes. Those are dog-wolf hybrids. / *Trust* me.

Islander H But anyway this is all neither here nor there.

Islander A Half-dog, half-wolf

Islander H The point / is . . .

Islander E Right the point is

Islander B . . .

Islander H But um . . .

Islander B The whales.

Islander D Yes.

Islander E Here we go.

Islander F The whales. For many years

Islander C So my great-great-grandmother was one of the original settlers here, and her name was

Islander G We don't need her whole backstory

Islander H (There *were* already people here)

Islander C I'm just saying, her name was

Islander H (There were Native Americans here)

Islander D Just let her say the name

Islander C EDNA ST. VINCENT MILLAY.

Islander H (I just want people to know that *we* know that there were Native Americans here)

Islander C But not the one you've heard of: different one. Different Edna St. Vincent Millay. They happened to have the same name. But my great-great-grandmother was quite a trailblazer in her own right. She came over in a boat, she was pulled by the Native Americans, and she wrote about it in her journal, she was a diarist, she was a great diarist

Islander F She was a great chronicler

Islander C She was *the* great chronicler, the great diarist

Islander B Well . . .

Islander C Of the island

Islander B Right.

Islander C And what made a *particular impression on her*, living here in the 1800s, was that if you stood on the beach at just the right time, you could often see a family of whales, of orca whales (Do you know them? Black, with the . . . white . . .)

Islander G People know orcas

Islander C I'm just checking! . . . moving through the water, and you could see their fins, moving in strange patterns, oh and certain members were born or would die, but this family of orcas had come every year for generations, up until the current day. And sometimes they'd leap!

Islander F They'd leap into the air, at dusk

Islander E The most beautiful thing

Islander D You'd see it, out of the corner of your eye

Islander B If you were driving by the water

Islander F As if in slow motion

Islander A Emerging . . . rising . . . cresting . . . falling

Islander E "Did I see that?"

Islander D Splash!

Islander H ("Did *I* doooo thaaaat"—is that a quote from something?)

Islander G And then another, and another

Islander A Emerging . . . rising . . . cresting . . . falling . . . **Islander E** (Some TV show)

Islander B Whales

Islander D Our whales **Islander E** (Jim Carrey?)

Islander F Our beautiful whales

Islander A In the water around our island. **Islander H** (No . . .)

Islander G But something was different this year.

Islander F I would often go down to the beach, and look over the water, at night, holding a glass of white wine

Islander D And I'd do the same, with mint tea because I *don't* drink

Islander B And I'd drive with my mom, to the beach, we don't have beachfront property

Islander G There's a slight division here, between those who don't and those who do

Islander C A *huge* division

Islander E ("Did I do that . . .")

Islander H ("Somebody *shoot* me": *that's* Jim Carrey)

Islander C I've worked at Thriftway, for *thirty years*. I've rung up your groceries I don't *know* how many times. And do you *ever* ask about my back?

Islander D And I'd stand on the edge of the ferry boat, after a long day in the city, and we'd all look out, over that still water:

Islander B Nothing

Islander E Why weren't they coming

Islander C They had disappeared.

Islander H And I was upset

Islander E Because our island without the whales

Islander A It's still beautiful

Islander G Without the whales it's still beautiful but it's

Islander F Little prayers, I'd sometimes say, walking down, to the edge of the water, my arms raised, as the sun fell, to the whales, for protection against harm

Islander H When my son got really sick

Islander C It was said that if you came down to the water, and asked them, that they'd

Islander H And it was true, I asked them and after that he was

Islander F Did they need *us*?

Islander D Were they sick somehow?

Islander B Maybe they were teasing us

Islander E "Ya dirty bastards!"

Islander H (Is *that* from a movie?)

Islander E (No)

Islander A Point was: they'd gone missing

Islander F And it was unacceptable

Islander C And we'd have to find them

Islander B As a community

Islander G We'd have to come together

Islander A In a common mission

Islander D To find the whales

Islander E Because without them

Islander G The empty ocean

Islander B (There'd be fish, sure)

Islander A No whales?

Islander D They've cared for us, I think, over the years

Islander B As best they could

Islander A And now it's our job

Islander E To care for *them*.

Islander F I mean, what the hell?!

Mayor Annie WE JUST HAVE TO FIND THOSE WHALES.

Scene Two

John It's very "of the 90s"

Mary Yes

John "Save the whales"

Mary Yes

John It's almost quaint, isn't it?

Mary But now it's *find* the whales

John It's different, isn't it: the *finding* is more about *us*

Mary Maybe they don't want to be found
Maybe they're somewhere else

John The great blue beyond . . .

Mary (*re: a sad little rack of sunglasses*) What do you think?

John We might need to go off-island for sunglasses

Mary Or order them on—( )

John Don't say it! I like to shop local

Mary The pharmacy has such a limited selection

John It's nice to have limits

Karl You folks finding what you need?

John Oh hey there, Karl

Karl Just another day in paradise **Mary** You got it! / Oh, what?

John Awesome. Doin' good.

Karl Hey let me know if you need me to order anything, anything we don't have we can order . . . just because you don't see it, doesn't mean we can't get it. Odds are? We can.

Mary Fantastic! **John** Cheers! That's great to know

Karl *leaves.*

Mary (It's so sad)

John (Shhh) Bill Clinton

Mary Hmmm?

John 90s stuff . . .
(*Re: sunglasses*) Ooooh

Mary (*re: sunglasses*) Eh

John The primary boy bands, and the secondary boy bands . . .

Mary What was the one with Nick um

John And then the tertiary boy bands, whose names are lost to history . . .

Mary Save the rainforest was very 90s, too

John As a movement? / (Le-shay?)

Mary As something on our minds

John I think that was more 80s

Mary Was it?

John No 80s was hole in the ozone

Mary No that was very early 90s, I think

John Ah

Mary And then we solved that, and we moved on to the rainforest

John And did we solve *that*? / (Lah-shay.)

Mary I don't think so . . . global warming sort of came to encompass the rainforest, and the whales

John Well, then, it seems very retro doesn't it? To pick one species and be concerned

Mary I mean . . . *are* you concerned? Personally?

John I love those whales! Don't you?

Mary Sure. But doesn't it seem like maybe they've just . . . gone somewhere else for a while?

John Could be

Mary People get too scared about things, everyone's so nervous

John On the other hand . . . there's a lot to be nervous about

Mary Jodi's been trying to get me to come to Quaker Meeting . . .

John I went once, I thought that might be the answer

Mary And?

John You just *sit* there

Mary But you're in chairs, right? Zen dest*royed* my knees.

John You're in chairs, yes, but people are *constantly* sighing, there's half an hour of sighing followed by thirty minutes of looking at the clock

Mary Hah, well. So that's not the answer.

John Sidebar: were you living here when the county tried to build a bridge to connect the island to the mainland, to replace the ferries

Mary Good God

John I chained myself

Mary You did not

John I did! I chained myself to the thing. To the whatever. To stop it.
I had a different internal sense of who I was . . .
I'm getting more and more confused as I get older about who the hell I actually am.
All of a sudden I'm thinking am I man who wears jewelry? My whole life my fingers have been unadorned, *as* has my neck, but *now* I'm thinking I want to do a sort of pirate thing, like I want my hands to flash when they move through the air

Mary I support that, John

John But no it was actually a blast: it was me, and Brock Taylor and his son who did *not* want to be there, and the Benedictine monks

Mary This was before my time.

John We didn't know each other yet! Sad

Mary I was still living in Seattle, still pursuing the dream

John Which dream was that?

Mary Massage school . . .

John Was Chris in the picture

Mary *Can we not talk about Chris?*

John Sorry

Mary Sorry.
He keeps calling me, he keeps showing up! I said don't show up unless it's your day to take Carrie, but he wants to talk, I said I don't want to talk, and then he's positioning himself like he's pro-dialogue and I'm anti-dialogue, the fucker, when it's like no I just don't want to be with someone who is physically uh aggressive towards me. That's a clear boundary for me.

John Mmm

Mary Subject change because I know what you're thinking

John Nothing you want to—?

Mary No. I can hear your thoughts.

John Mkay.
Shall we head out?

Mary I'll just get these, what the hell

Karl You found something you like!

Mary Love these!

Karl We try to have stuff that people want

John There's no better place than the pharmacy

Karl We love to hear it!
I'll just bag these up

Mary I'll wear 'em!
(*To* **John**) I have to pick up Carrie soon but I was thinking of grabbing groceries briefly can you stand it

John I'll come I need dinner stuff hey have you seen Ella?

Mary I was supposed to get together with her but she canceled

John Yeah I've been trying to reach her but she's not responding she's

Mary Maybe she's busy with her daughter she's

John I feel like ever since the chemo she's been keeping her distance it's

Mary I'm worried about her I want to know what the doctors

John We should call her together we should

Mary Yeah maybe if we contact her together it'll

Karl Now you take care now!

*

Mayor Annie Editorial from the Mayor, from Mayor Annie, for this week's paper . . . In this moment of . . . when everything's, ah . . . in this difficult moment, when we, ah . . . it's just *crucial* that we . . .
I can't think of what to say!

*

Karl Fucking endless supply of sunglasses. We have as *many* sunglasses as I can *afford* to *have* in *stock.*

*

Chris Hi.

Mary What. Hi.

Chris Fancy running into you here

Mary Imagine the chances.

Chris Browsing for soup

Mary Yeah well it's cold out and I'm seeking comfort.

Chris Can we talk

Mary About what

Chris I want to see Carrie more

Mary It's not in the agreement

Chris I know it's not in the agreement but I want to see her more

Mary You're not kind to her

Chris That's not true

Mary The things she tells me

Chris I set *rules*

Mary Oh okay

Chris And she interprets that as not being kind

Mary Don't tell *me* I *don't* set rules

Chris I didn't say that

Mary You said you set rules

Chris I do set rules

Mary Which implies that I don't set rules

Chris No I didn't imply that, I said I set rules, I'm not speaking to whether you set rules

Mary Just saying that implies that you think I don't set rules

Chris Name one rule that you set, and enforced, around the house.

Mary "Respect."

Chris Well, that's just incredibly vague.

Mary The rules weren't the issue, it wasn't an issue of the rules. It was an issue of you and me.

Chris I want to work on it, I told you that.

Mary Yes, well

Chris I think we can *work* on our communication

Mary It's not about our communication it's about your explosive, frightening / anger

Chris Can I come by? Sometime? To see you? To / talk more

Mary No. I don't know. Maybe. Not right now.

John *walks up.*

John I found my tempeh

Chris Hello, John

John Chris

Mary . . .

John . . .

Chris . . .

*

Karl Do you want an empty downtown broken windows graffiti everywhere total desolation or do you want to SUPPORT LOCAL BUSINESSES

*

John Through a pen pal service

Ella What is a pen pal service

Mary Oh my God she is so DUMB

John She's not dumb, she's—

Mary Head in the clouds

John (Well, she's a little dumb) **Ella** That's a nice way of putting it

John So she met him on this service that's um, they set you up with pen pals, based on your common interests

Mary What are her common interests?

Ella Horses?

Mary Every time I see her she goes on and on about those horses

Ella Horses to me are very boring

Mary I know!

John Apparently she's had different pen pals over the years, some of them romantic, and she writes to them by hand which is sort of I think sweet

Mary Does she send a picture?

John Stop it!

Mary I'm just saying

John I know what you're saying.

Ella Maybe she sends a picture of her horses

John They have the same issue.

Ella Oh dear.

Mary You comfortable, Ella?

Ella Yes I'm fine stop fussing, and both of you need to stop looking at me like that

Mary Like what

Ella With concern I told you I'm feeling much better

Mary Okay sorry

Ella I just want to not talk about all that now, it was horrible and embarrassing and painful and *boring* and it's OVER and I want to just be here with you both and gossip. That's what I want.

Little beat.

John Alright then back to making fun of Leslie. As far as I know she's never met any of these men. They just have these protracted correspondences. Correspondence—

Mary -es. Correspondences?

Ella I actually took a writing class with her, with Leslie

Mary Oh did you?

John (They correspond.)

Ella I did, at the library, a poetry class, it was

Mary How is her poetry?

Ella Well, it's not good

John Oh no

Ella It's very actually terrible

John Well, I'm not surprised

Ella But you know what? *Everyone*'s was terrible

John Ha! Well, sure

Ella *Mine* was terrible.

Mary Oh stop it

Ella (*to* **Mary**) I read you some of it!

Mary It wasn't bad!

Ella We both agreed it was bad

Mary You were just starting out

Ella Everyone's was bad. Which took the pressure off.
Except for Joy Mead, you know her?

Mary Joy Mead . . .

John She edits the paper

Ella She brought in a poem and you know we'd sat through all these other poems, that were very terrible, when I'm there I'm just thinking thank *God* one of you isn't there or I would die of the giggles, but then Joy Mead reads hers and I thought—this is a *real* poem. Which ruined the class, actually. It was a fascinating poem. It was about being—I think—in a cult?

Mary I don't know her.

John Yes you do **Ella** You do

John She writes that column for the Beachcomber? "Talking Island"?

Ella I think it's "Island Talking"

John I think it's like "Talkin' Island"

Ella She's actually . . . I'm working with her on something. With Joy Mead.

John Oh on what?

Ella Oh just on a project . . .

Mary . . . Okay . . . . **John** Mysterious!

Mary So they really say you won't need any more treatment?

Ella That's what they say, yeah

John Well, that's wonderful **Mary** God that's good, Ella

Ella Yes! I'm very relieved.
Tests every six months, but other than that: done.

Mary Wow.

Ella Yeah.

John Thank God. Or whoever.

Ella Right.

Mary Knock on wood.

Ella I say a little prayer to the forest sometimes, or the water.

Mary I like that.

John So . . . life goes on?

Ella Life goes on.

She runs her finger around the rim of her glass.

*

Joy Mead It's an interesting project for me, I've never collaborated on an obituary with the person who might soon die. Usually we get the notice, that someone has died, and then I write them, or the family writes them . . . but it's wonderful that you

Ella Well, my daughter thinks me doing it myself is weird. But she thinks everything I do is weird or bad or

Joy Mead That's having kids for you

Ella Exactly! Anyway the nice thing about this project is that there is a clear deadline! A very firm deadline. (I'm doing an assisted sort of)

Joy Mead Oh you're having it

Ella Having it done, yes. Legally. They're sending a nurse to help.
I'm sort of—not telling anyone right now though? Don't tell anyone. If you don't mind.

Joy Mead Of course.

Ella It's wonderful that new law, that we can just

Joy Mead But you look so well!

Ella I know. But this thing is spreading inside me. You wouldn't know it. Sometimes *I* don't believe it. And at the moment I feel okay! It's I guess this brief window. I've emerged from several *brutal* rounds of chemotherapy

Joy Mead Oh goodness

Ella Which was very nightmarish and you know? Didn't actually work. So much for modern medicine.

Joy Mead Should have stuck to the supplements.

Ella Yes! Should have done the goat milk technique.

Joy Mead I had an aunt who had cancer, forgive me if this story is

Ella No I love the stories about people who have had cancer, truly

Joy Mead Good, well, my aunt had terrible, you know metastatic sort of and she *refused* treatment she said I'm going to think positive thoughts! And I'm going to eat *completely* vegan. Not a *thing* that has been produced cruelly will touch my lips.

Ella And did it work?

Joy Mead Briefly.

Ella Right. I've been resolutely avoiding positive thinking. I figure why start now.

Joy Mead Ha!

Ella Anyway. Thank you—

Joy Mead Yes, shall we / begin?

Ella —and I have to tell you that ever since we took that class

Joy Mead Oh goodness I am not a poet

Ella Oh my God. That poem you brought in about moving to the island and taking the ferry boat across the ocean from the city for the first time with the wind in your face at night and the phosphorescence in the water and the connection you made between that and your *divorce*?
I still think about that poem.

Joy Mead Well, thank you.
I thought *your* poetry was—

Ella No.

Joy Mead Okay.

Ella Thank you, but.

Joy Mead Well. Shall we get started then?

Ella Yes. God. How do we

Joy Mead Why don't you tell me where you

Ella Oh where I—?
. . .
No but why don't you tell me a little about you first. I mean how did you get into reporting? What a fascinating / profession

*

Mayor Annie Rules

Islander A Rules?

Mayor Annie We need rules, for speaking, for who speaks when, if everyone's going to show up, we need a moderator. I'll moderate, obviously, as mayor. How many people do you think will show up. A hundred? Hundred and fifty?

Islander A I think we should plan for ten

Mayor Annie There are more than ten people on this island who care about our whales

Islander A People get busy

Mayor Annie This is the whales we're talking about here!

Islander A Alright, let's plan for twenty. In the past everyone has voted to elect a moderator at the beginning of the meeting.

Mayor Annie Well, I think that system in the past has led to chaos. Bad people get elected moderator. Stupid people.

Islander A Happens every day . . .

Mayor Annie Ha ha. But okay. So I'll moderate. Obviously if there's a community meeting, the mayor should moderate.

Islander A Well, I don't know that that *is* obvious

Mayor Annie Well, what's the point of a mayor then?

Islander A Well, I don't know that there *is* a point

Mayor Annie That's not my point, my point is

Islander A You're a ceremonial mayor elected because people like dogs

Mayor Annie And goddamnit, I'm gonna make use of my time in office. To do some goddamn good. I mean no one else is stepping up to the plate to figure out this goddamn whale situation, I mean we've got to get some scientists out here—Joe's a scientist right?

Islander A He was

Mayor Annie What do you mean *was*

Islander A Joe moved to Minnesota about eight years ago

Mayor Annie Well, that's why I haven't seen him!

Islander A You have to pay attention. You have to be more aware of everyone around you. I've been trying to tell you that for years.

Mayor Annie You have?

Islander A You're wrapped up in your own head. You can't see the people around you. You have to improve on that. If you're going to be a good mayor.

Mayor Annie Well

Islander A Yeah

Mayor Annie That's good, no, thank you for that I will work on / that

*

Joy Mead Involved very briefly in um well I guess I have to call it a cult

Ella That's fascinating

Joy Mead Although of course when you're in the middle of a cult you don't call it that

Ella Of course not!

Joy Mead Have you ever been part of a cult?

Ella Well, let's see. There was a period in my life where I was going to Curves every day, you know the, the . . . . women's gym?

Joy Mead . . .

Ella But that wasn't really a cult **Joy Mead** Sounds like more of a hobby

Ella That was a hobby, yes, so *no*, I suppose, I haven't . . .

Joy Mead So I was living in a house alone, I had just gotten divorced, and it wasn't the whole cult, it was just a family, who were members of the cult, and every weekend they'd spend on the compound in the woods, where the actual cult was located, but during the week they were just living their normal lives as a family, so I lived with them in this house in Seattle, I had the bottom floor and they had the top. And they were *very* sweet people.

Ella Well, did they ever try to recruit you?

Joy Mead Well, that's the strange thing they really *didn't*, and I was very on guard against it, but the fact that they didn't seem interested in having me as a member, made me kind of interested. So one weekend finally, they invited me to come out, to the compound

Ella Oh this is fascinating

Joy Mead We get there and the leader walked right towards me. I'd seen photos of him and always thought what a dumpy white man who'd taken this Indian name you know, but he walked towards me and there was indeed some powerful *thing* around this person, and I thought uh-oh. Even me, skeptic: he was magnetic. There was a whole meal, and everyone was—their faces were shining and they sat me right next to the leader and there was something alive in the room that in spite of every skeptical bone in my body I felt something inside me start to—open.

Ella How was the food?

Joy Mead Terrible! This massive communal lentil loaf

Ella Good God

Joy Mead And then these strange vegetables. I asked someone what *are* these vegetables? Never got an answer.

*

Ali Being weird about it

Ella I'm not being weird about it, I've just gone and had my obituary done, well we didn't get to it because we started talking about other things—have you ever talked to Joy Mead? She's a *fascinating* / person

Ali I find her column to be very boring

Ella Boring? No, it's

Ali It's very local

Ella Well, it's a local paper

Ali Yes but even within the context of a local paper, the issues she covers are very / local

Ella Okay, well, I mean for those of us who *live* here, it's nice to have things be local. I know it's not *New York*

Ali Don't do that

Ella Listen I understand what you're saying. But she's a big city reporter at heart—she was in a cult!

Ali Well, good. I think it's good that you're doing this.

Ella You do?

Ali Yeah, I do. I mean I offered to sit down with you and write about your life, I think that could be really / nice

Ella No no, it's better this way. It's easier to do with someone else.
But I agree that it's good. I think it'll help . . . making everything real. I mean I know it's real. But then other times I keep thinking: maybe they made a mistake, mixed up the scans

Ali Uhhhh

Ella These things do happen!

*

Mo Alexander?

Alexander What, Mom?

Mo We're having dinner

Alexander I'm not hungry

Mo I'd like you to come down

Alexander I'm practicing for my dance recital

Mo I can see that

Alexander It has to be perfect

Mo I know, honey, but you have to take breaks, for meals

Alexander Dance is my priority

Mo I know it is *now* but

Alexander Dance will *always* be my priority. I'm going to be a dancer.

Mo Well . . .

Alexander I'm *going* to be a dancer, Mom. I'm good.

Mo Why don't you show me a little bit then.

Alexander No! It's not ready.
Okay fine. I'll show you just a tiny part of it.

He dances.

Alexander There. Mom? What did you think.
Tell me for real.

Mo It's . . . it's very nice.

Alexander Well it's not *ready*, I *told* you it wasn't / good yet!

*

Ali Need to have some kind of plan for it, like for how you want it to feel on the day.

Ella I don't know

Ali Or any, yeah, ritual or

Ella I don't want a ritual.

Ali There's nothing that you want to like—celebrate your life?

Ella Well, I want fresh flowers

Ali Why don't we ask John to bring flowers from his garden.

Ella I don't want John involved.

Ali Have you told him that you're going ahead with it?
Mom.
Have you even told him you're still sick?

Ella . . . not yet

Ali He's your best friend.

Ella I know that. You think I don't know that?

Ali He would want to be there. You have to tell him. Mom!

Ella Just leave it, okay? I need you to stop pressuring me. I need you to back off.
. . .
I want you to take care of yourself through all this. Have you been keeping in touch with your friends?

Ali Yes

Ella Any word from Shaina?

Ali I told you we're on a break.

Ella I *really* like her.

Ali You've made that very clear.

Ella She seems to really care about you and be up for you know your *moods*

Ali I told you we are *on* a break.

Ella Okay okay sorry
I'll try to be more proactive about planning my big day.
There is someone else I'd like to come actually.

Ali Who?

Ella I have to ask her first.

*

John And I'm John

Gary Okay

John So yes. That's me and I'd like to hire you to uh

Gary Chop the wood

John Well, yes, actually, so uh—well, first, are you hungry?

Gary No I ate

John You ate

Gary Yes

John Seen you walking around with that chainsaw for many years, and I have all this wood, the storm, you know the storm?

Gary No

John There was a pretty big storm?

Gary No

John Okay well it brought down a lot of trees

Gary Seen a lot of trees down

John Yes! Probly from the storm.

Gary Which made me kinda think that more people would be hiring me and stuff lately to chop but no go

John Okay

Gary Yeah no go on that, keep walking around but no one's stopped and picked me up brought me back to their place until you so—no go, yeah, no / go

John Well, that's strange you'd think there'd be lots of offers

Gary So I do appreciate the work

John You're sure you're not hungry?

Gary No I just ate about a minute or two ago

John . . .
Listen I've always wondered, where do you stay? At night, I mean, do you live with someone, I know there's no ah shelters on the island

Gary I don't live in a shelter

John No I know

Gary How come you're saying that?

John Well, I don't know, I mean . . . aren't you / homeless

Gary I have a home

John Okay, yeah, great! Well, that's wonderful.

Gary How come you're saying that maybe I'm homeless?

John I mean your name is Homeless Gary

Gary That's not my name

John I guess that's what people call you, that's . . . were you not aware of that?

Gary My name is Gary. Gary Williams.

John Of course. I'm sorry. I didn't . . . I'm sorry . . . .
Let's go take a look at the wood out back

Gary I gotta grease up my chainsaw

John Okay, well, maybe I should show you the / wood

Gary I gotta grease it up out front the chainsaw get it all greased up and then we can take a look at the wood out back

John Okay . . . well . . . great . . .

*

Mayor Annie Editorial from the Mayor. Ummmmmm . . . .

*

Mary What the fuck are you doing here?

Chris I don't know I was just driving by

Mary I specifically said don't come here this is not charming behavior this may have been charming fifteen or even ten years ago but things have changed, culturally this is considered abusive

Chris Is Carrie here?

Mary It's not your day, no, she's at a friend's house

Chris What friend?

Mary Why are you here?

Chris What friend?

Mary Ainsley

Chris I don't want her hanging out with Ainsley, her mom's a meth addict

Mary Yeah, well

Chris So I don't want her hanging out with her

Mary She likes Ainsley. And I'll let her hang out with Ainsley if she fucking wants.

Chris . . .

Mary . . .

Chris Garden looks good

Mary It's winter there's nothing growing

Chris Yeah, well, you've wrapped everything up, you've wrapped the bushes up just like I used to . . . I appreciate that. Taking care of them for the winter. That's thoughtful. That shows some real care and concern for the bushes. You kinda tied this one wrong.

Mary . . .

Chris But that's okay

Mary Why did you come here, what do you want . . .

Chris I want to try and be different

Mary Jesus

Chris No really though

Mary I know that's what you want! I know you want it very sincerely, every time you want it. You *really* do.

Chris I miss you really badly. I miss Carrie like I'm in pain. I feel sick.
I told you fifty times I'm really sorry. I fucking said it over and over, what else do you want?
I'm sorry.
I need help.
I swear to you I can be different.

*

Leslie (*deep sigh*) I am writing you this letter . . . after a *difficult day* with my horses. I won't go into it in *great* detail but

*

Mayor Annie Editorial from the Mayor. I so appreciate all of you who have taken the time to send your congratulations, to me, about me becoming Mayor. I must stress that I am aware that this is a ceremonial position, but in a time like now where uh—where things are going on, I must say that I intend not to waste my time in this position, and I intend to do some good. I think we've all just got to get together and see if we can get some information about those whales. Later we'll deal with the stray dog problem but at the moment they don't seem to be causing any issues and I am just much more concerned about those whales. So *please* get involved and stay involved. In closing, the holidays are coming up, so happy holidays, no matter *what* you celebrate, or whether you're just observing the change of seasons. Sincerely, "Mayor Annie." P.S. I hope some of you can make it to the meeting to discuss the whales.

*

It begins to snow.

Islanders (*multiple voices, sung*) God bless you merry, gentlemen

Everyone sings the song but there aren't enough lyric sheets to go around and only some of the people know the words.

There's a lot of humming and sort of trying to hear what other people are saying.

But it finishes strong.

Islander F That's *very* good. *Very* good. You've all been practicing at home, and it *shows*.

Islander B You sounded great tonight.

Islander C Oh did I? Well, I thought you sounded very nice as well.

Islander D (Been hanging in there?)

Islander A (Yeah! Doing alright, doing alright . . .)

Islander D (You should come over some time, I'll make dinner)

Islander A (Sure, yeah, no, that'd be fun, no definitely)

Islander G Anyone had a whale sighting?

Islander B No

Islander H Sadly, no

Islander E Not me

Islander A Can't say I have

Islander G (Winter came so fast)

Islander H (I know. I barely registered the fall.)

Scene Three

Ella *is onstage, alone, in the near-dark.*

Voice of a Therapist What do you see.

Ella I'm on a boat . . .

Voice of a Therapist Is there anyone in the boat with you?

Ella I don't think so . . .
I'm alone.
There's a paddle in my hand . . . I'm going somewhere, but I don't know where
The island's in the distance, getting farther and farther away and
I don't know where I'm going

Voice of a Therapist Say more.

Ella It's not bad to be out on the water
It's nice to be away from everything, I have always dreamed of being completely unreachable.
But I do miss my daughter.
I feel a clear feeling of—missing her.
Do you mind if we stop now? I'm tired.

Voice of a Therapist Of course.

Ella So uh . . . I'll see you next week. If I'm feeling up to it. I think things are about to turn.

Voice of a Therapist Why do you say that?

Ella I just feel it coming.

Voice of a Therapist Ella.

Ella What.

Voice of a Therapist It's okay to be scared.

Ella I know that.

Voice of a Therapist It might be helpful to say it.

Ella Say what.

Voice of a Therapist . . .

Ella I am . . . scared.

Voice of a Therapist Say more?

Ella I am angry

Voice of a Therapist Yes.

Ella I am angry at myself.
I am so angry.
I have *wasted my fucking life.*

A flock of birds flies across the sky—overwhelming and fast—then they are gone.

Scene Four

Islander H Migratory patterns being such that

Islander A Is that really the right um

Islander H Yes

Islander A Word, migratory

Islander H Or whatever, patterns, such that, we can tell

Islander F Very old school isn't it (the map)

Islander E I like it, kinda vintage

Islander H Such that we can tell that

Islander G And I've really looked into this online, so I'm happy to confirm

Islander H Oh okay would you like to

Islander G No, sorry

Islander H Great so anyway

Islander B Yes let's

Islander H *I* can actually confirm that

Islander C Do you want the

Islander H No no I'll do it without the

Islander B Would you like me to

Islander H Only if you'd

Islander B I find that it can

Islander H Sure then. Sure. I find that the

Islander A Talk about the seasons

Islander F We know about the seasons

Islander A Talk about the currents then

Islander B We'll get to the currents

Islander H We're just doing a general overview of the

Islander F Okay lemme

Islander H Here is the West Coast. Of the United States. So the whales begin here, when the weather is warm, all the way up by Alaska, and then they come down, following currents and temperatures—nearly the whole height of the world, from top to bottom, eating fish all the way, and sleeping on one side of their body at a time—which is amazing—(*demonstrating:*) sleep/awake, sleep/awake—and then they end up here. In Mexico. And then they come back for a while and hang out outside of San Diego. And then when August is kind of turning into September as I understand it they begin to make their way back all the way up here, to the Pacific Northwest, and they arrive here just as fall is ending, just as it's getting cold. Here in the water around our island. At least this pod does. And that's when we see them jump

Islander C Which is my favorite thing

Islander A I love going down to the water just as twilight is setting

Islander D Coming?

Islander A Twilight is coming, the sun is setting

Islander G Golden

Islander D Or sometimes in the morning, just when the sun is

Islander C Spreading its um its fingers

Islander F *Love* that language yes in those moments when the day has um before it's started or when it's about to end

Islander B And the light is kind of

Islander D And everything's just

Islander B Exactly

Islander D Suspended and still

Islander C The way sometimes in the water there will be nothing, nothing . . . and then they'll JUMP

The **Islanders** *imagine it.*

Islander H Anyway. So that's the situation as it stands now. And I know all of you need to get home to your kids and to dinner if you have kids and just to dinner if you don't have kids. So now that we're all on the same page, maybe next time let's focus on developing a plan of action to find them. Sound good?

Islander C Go team!

*

John Getting dark

Gary Yeah
But I'm not done with the wood

John I get it
But you've done a really good effort and I think maybe you could stop now

Gary Yeah, but I'm not done

John No I get that but I hired you and I'm telling you it's okay

Gary I can't stop if I'm not done. You hired me to do a job

John No yes but listen I'll pay you the full amount it's just

Gary Do you not want your wood chopped?

John I do! I do but it's just that this is plenty of wood and it's I'm making some dinner and I was wondering if you want to

Gary Oh

John You're welcome to, if you'd like, I'm going to

Gary I have to finish this wood.

John Oh, well, alright.

Gary . . .

John . . .

Gary Chop chop

John Making of the dinner

*

Mo Alexander?

Alexander Yes

Mo You have to come down for dinner

Alexander I'm dancing, watch me

Mo I can't right now

Alexander Watch me I'm getting better, right? Watch me for a second

Mo Honey

Alexander For a second

Mo Okay, for one second

He dances.

Alexander I'm getting better right

Mo I don't know, to be honest

Alexander Mom!

Mo What matters is you're having fun but I want you to keep up with your schoolwork

Alexander I don't care about my schoolwork

Mo That's what I'm concerned about

*

Joy Mead *and* **Ella** *have dinner. Delicious, comfortable. A candle.*

Joy Mead There was something really wonderful about the community of the cult: to be connected to God *all of the time* during every moment of the day

Ella I can't imagine, it sounds wonderful

Joy Mead I mean I've tried to do that my whole life feeling like every moment matters or whatever but up until then I'd failed

Ella It's hard to do

Joy Mead My mind is always racing, regretting the past or imagining different kinds of lives I could have had or

Ella Me too!

Joy Mead There was something about that environment where I was able to just—settle. It could have been the natural beauty, you had to drive a very long road to get to the main building, the fellowship hall, the sides of the road were strewn with wildflowers, the drive was truly endless and then you emerged to see this beautiful green field and beyond it the vast blue ocean, sound of wind and birds: two women in robes came out to meet me with slices of fruit.

Ella Incredible
Would you like some more?

Joy Mead No but this is a fantastic

Ella This is my special

Joy Mead And this salad!

Ella These are the first greens of spring

Joy Mead Thank you for having me over by the way

Ella Oh I'm so glad you said yes!

Joy Mead And at some point we have to get to that obituary

Ella Yes but tell me more. First. About the cult. So you get there

Joy Mead I was so entranced by the place that I didn't actually go home for weeks. I stayed there for several weeks. We had a very strict schedule I always knew where I had to be and when. At 6 a.m. we prayed, 7 a.m. a simple breakfast. Work for four hours, more prayer. I felt like a different person: amidst all the work I was at rest. There *were* weird things about it too I mean having sex with Robert (the cult leader, I could *not* call him by his Indian name even when I was deep in it) was interesting he had the smallest penis

Ella Oh God!

Joy Mead I've ever seen in my life

Ella That's wonderful, I mean as a detail, that's a funny detail

Joy Mead Really microscopic a um—how do you say it

Ella A micropenis

Joy Mead Yes, almost invisible, there amongst the—voluminous

Ella Stop it!

Joy Mead Waves of his pubic hair, really a *very* full bush

Ella You are too much!

Joy Mead And we were supposed to worship it! Not all of the time, it wasn't

Ella Sure

Joy Mead A phallic cult and he was very careful to worship our vaginas

Ella Naturally

Joy Mead I mean me and whoever else he was sleeping with

Ella Were there threesomes?

Joy Mead No no no hahaha it wasn't *that* bad.
Have you ever had a threesome?

Ella Who me? No no. I never did.

*

Leslie (*talking to her pen pal*) Nice to talk to you on the phone I've very much been enjoying writing back and forth, by post, it's rare, isn't it? Rare and wonderful. Oh, you know, I just finished feeding my horses and now I'm making dinner for myself. I like to snack for dinner, I like to pretend I'm having just a bunch of little snacks, as if I'm in Spain . . . well, that's how I imagine they do things in Spain, have you been?

No me neither! That's wonderful well um. . . . well we'll have to go! Maybe we could meet for the first time in Spain . . . if you want.
Listen I've written a new poem, and I'd love to read it to you, if you'd . . .
It's called "Life."

Joy Mead (*still at dinner with* **Ella**) And he leaned forward and whispered to me—"if you're very quiet, you might hear it."

Leslie Life.

Ella Hear it?

Joy Mead If you're very quiet, you *might* hear it.

Leslie Life is funny

Ella What is *it.*

Leslie Some say it's funny

Joy Mead *It.*

Leslie But as for me

Ella I don't understand.

Leslie Life has been a bitter fruit.

Joy Mead If you're very quiet, you might hear it.

Leslie With sweet seeds

Joy Mead If you're very *still*, you might *see* it.

Leslie Hard to find

Joy Mead You *might* hear it.

Leslie But *very* tasty

Joy Mead You *might* see it.

Leslie When you do.

Joy Mead It might be in the room.

Joy Mead *and* **Ella** *sit in silence—waiting to hear it.*

*

Mary Communication, okay! It's these sudden it's lashing out, suddenly, at me, as if you're

Chris Well, I get frustrated

Mary I know that I mean my God this whole marriage has been an immersion in your frustration—and things have been good, often they've been good, I love raising a daughter with you, loved

Chris I do too

Mary But don't you think it's hard on her, I mean you're so kind, when you're kind, but then the screaming

Chris I know. Don't you think I know? That it's a problem? I don't like—I mean I'm *aware* of it as a problem, don't you think I *know* / that?

Mary You're building to it now

Chris I'm not building to it

Mary Yes this is how it happens, don't you remember the last time that / it

Chris Yes of course I do

Mary Don't you see that that's a problem?

Chris Fine well, yeah well . . . I can see / that.

*

Ella Listen I uh this has been the most wonderful evening, thank you for coming

Joy Mead I hope hasn't been too tiring, to have me

Ella No! This is exactly what I need right now.
There is something uh—I need to ask you.

Joy Mead Uh-oh.
When someone says they need to ask me something

Ella Oh

Joy Mead Am I in trouble?

Ella Not like that! I um—I
I wanted to ask you if you'd—be in the room when I do it.

Joy Mead Oh.

Ella I know we've just met and this is . . . I know this is a big ask, but I feel so connected to you and I think us not knowing each other that well will be a good thing? You can just come, it'll only take an hour or two, and you can sit there, my daughter will be there too so I promise it won't be too awkward, and if you wanted to read something you could, or write a little something you could . . .

Joy Mead Oh.
Well . . .
I don't think so.

Ella You can think about it if you want.

Joy Mead No, I don't think so.

Ella Oh.
I thought maybe you would say yes.

Joy Mead . . . this has been wonderful, to get to know you, and it was such a pleasure to work on your obituary

Ella Well, we're not done

Joy Mead No

Ella We've barely even gotten to it, the conversation has been so wonderful, hasn't it?

Joy Mead It's been very nice . . . but, you know, it's mostly been me talking

Ella Oh

Joy Mead Sorry . . . I'm a very direct person

Ella Of course

Joy Mead And you know, I don't even know you really . . . don't you have people who know you?

Ella Of course I do, I have my daughter, and my friends, but

Joy Mead They might be more appropriate, don't you think?

Ella Of course

Joy Mead You should be with your loved ones right now. That does seem—that does seem important.

. . .

Thank you again for having me over though. This was truly an amazing meal.

*

Joy *leaves.* **Ella** *begins to clear the table.*

During the following sequence, she may change her clothes, or be helped out of her normal clothes and into a gown—transitioning from well to sick.

Mary Haven't heard from you

Ella Been busy

Mary With what

Ella Getting back to everything! After the treatments, getting back to my life, and my daughter's been here, so, from New York, so that's kept me busy

Mary Are you avoiding me, for some reason

Ella No! Not at all

Mary I just feel as if we had a very intense experience, all of us, taking care of you, through the chemo, and now you're

Ella You weren't taking care of me

Mary I drove you back and forth to the hospital

Ella And I really appreciated that

Mary But now I feel this . . . distance, it's very strange. . . . I mean what is going on? I've been texting you, calling, wanting to get together . . . John said he came by and you pretended not to be home?

Ella I just need a little space right now. I'm sorry.

Mary Ella.

Ella What.

Mary Are you still sick?

Ella I need some space.

*

Voice of a Therapist How are you feeling today, Ella?

Ella Incredibly energized. Incredibly wonderful

Voice of a Therapist That's interesting

Ella Yes. I mean there are problems, my pain is increasing, and I've had to go to the hospital a couple times

Voice of a Therapist Who's taken you?

Ella My daughter

Voice of a Therapist Do you have more of a support system than that?

Ella Not right now.

*

John Ella

Ella What

John What are you doing

Ella What do you mean

John Why don't you call me, you're sick, I want to be there for you

Ella I feel fine

John You have to ask for help

Ella Goddamn you. Why do you keep telling me I'm sick when I'm not. I'm not sick. I have so much energy, there are so many things I want to do. I want to clear out all my boxes, I want to empty out the house. I want to think back on my life and remember it and I want to go walk on the beach and I am praying—praying—that the whales come back so I can see them one more time. I'm seeing friends

John Mary says that she tried to come see you

Ella Other friends! Other friends I have a lot of people here who care about me who want to see me when I
Oh

John What

Ella Pain

He helps her sit down.

*

Chris It just washes over me, it's like this wave and I feel it rising and can't stop it.

Voice of a Therapist Can you remember the first time you felt like that?

Chris Not really. There was a lot of anger in my house when I was a kid. My dad used to um hit me. He hit my mom too, she got it the worst. I feel proud that I haven't turned out like that, that I've . . . stopped the cycle. At least with my kid. I've *never* hit my kid.

Voice of a Therapist Have you been in therapy previously?

Chris I don't believe much in therapy. I mean today has been good but

Voice of a Therapist Well it doesn't matter too much whether you believe or not. It matters if you keep showing up. Are you going to keep showing up?

Chris Sorry. I just had a memory. About my dad.

Voice of a Therapist Do you want to tell me about it?

Chris Uhhh
. . .
Yeah so my dad was not someone who I

*

Gary Been a strange life. Walking back and forth with my chainsaw, on the island.

John Well, for me too it's been a strange few years. I was living with a man: Santiago, and we had a little gallery, on the island? Just showing local island painters, mixed quality—some good ones, some I'd be happy to set on fire. I'd met Santiago about ten years prior, he was a little younger than me—Venezuelan by birth, very kind eyes, *very* attractive, and he actually came into my gallery, and at that *exact* same moment

*

Mayor Annie Can we have a meeting? Can we please have a meeting? Can we? Can we have a meeting.

Islander G We're busy, Annie.

Mayor Annie Don't you care about the whales?

Islander G We do care but it just seems . . . out of our hands.

Mayor Annie No. We can't think that way. That's lazy. No. We will step up to the plate, we will be responsible stewards of this island, for the next generation.

Islander D It's probably something to do with climate change

Islander H "To do," are you British?

Islander D I spent time in Europe, in college

Islander H Okay that's great

Islander E Probably picked up some of the affectations

Islander B I mean they're out there in the water, somewhere, / hopefully

Islander D If they're still alive

Islander E Oh stop it

Islander C They're still alive

Islander A Don't be a Pollyanna

Islander C I'm not being naive, I'm saying, what, a whole family of whales is just going to die?

Islander A Sure

Islander C From what would they be dead?

Islander E Disease

Islander B Maybe they got chopped up, by a boat

Islander D Maybe they got too warm from global warming, and they sunk

Islander A Or their food sources are gone

Mayor Annie Basically . . . we need to come up with a plan.

Islander D Can we do Roberts rules of order?

Islander E What's that?

Islander C I know what that is

Mayor Annie Well, a lot of us don't know what that is

Islander B I think we should just take turns

Islander C I think we should have what's known in tribal cultures as "a speaking stick"

Islander A Don't be an idiot

Islander F Hey! **Islander D** Harsh. No need to be harsh.

Islander B I think we should go around the circle, and everyone should have two minutes. Keep it simple.

Islander E Can I pass?

Islander D Of course

Mayor Annie No. You can't pass. Everyone must participate.

Islander E Well, I don't know if I have an opinion

Mayor Annie Well, form one. Form an opinion. If you want, you can go last and listen to other people's opinions and pick one. That's fine. But it's your duty to have an opinion. Otherwise why are you here

Islander E I wanted to be a part of it

Islander B Okay I'm going to speak. What I think is

Islander G (And we developed a plan of action.)

Islander B Okay so

Islander D My sense—(I was on the phone with an oceanographer, he was from Germany although his accent was so light you could barely tell with whom I'd had a brief relationship when we were both in our twenties living in Monterey California at the time have you been there? Wonderful aquarium)—my sense is that there are larger forces at play

Voice of an Oceanographer Well, obviously there are larger forces at play

Islander D Such as global warming, my sense is that there are temperature changes

Voice of an Oceanographer Well, I'm aware of that I'm an oceanographer

Islander D And I know you're aware of that, I'm not trying to / condescend

Voice of an Oceanographer You're explaining global warming to / me

Islander D I'm just giving you my sense of things! Now listen: why isn't the whole oceanography community down here

Voice of an Oceanographer Because there are systemic disruptions, it's happening all over . . . I mean my God we've lost nearly all of the minnows, shrimp are nearly kaput, except for the ones in factory farms

Islander D Shrimp fields

Voice of an Oceanographer Exactly

Islander D Heard a podcast about that

Voice of an Oceanographer To say nothing of the dolphins, porpoises are just about kaput

Islander D Well, I know that

Voice of an Oceanographer What we're seeing here are truly crazy disruptions, species are behaving in ways that could disrupt vital food chains and send the whole ocean spiraling towards collapse

Islander D Well, why isn't anyone covering this, why isn't anyone sounding the alarm

Voice of an Oceanographer They have been! People have been sounding the alarm for years! I mean my God they've been sounding the alarm for years

Islander D Well, they should have said it louder they should have been more insistent and stuff!

Islander D *storms off.*

*

Islander G Dream of ah of being ah a whale, of one of our whales, funny to think about it / that way

Islander F No but I do, I do too, / our whales

Islander G It's funny to think about it that / way

Islander E No but I do too

Islander G Do you?

Islander E Yes I feel a sort of, not / fatherly but

Islander F I feel a very motherly sort of

Islander B I want to hug them. Can you hug a whale?

Islander G I think it would be challenging

Islander A They'd be slippery

Islander B They're very large

Islander B If you could somehow sort of secure them in place

Islander G Of course you'd have to ask for consent

Islander A You would wouldn't you

Islander E How would you do that

Islander G Can people speak whale language?

Islander B There is a language, I don't know / if

Islander G Aren't there words and things, in whale language

Islander F There *are* words

Islander B No

Islander F There are I studied it

Islander B No come on

Islander F I took a very bad class on—this is embarrassing but—sonic meditation?

Islander E What's that

Islander F It's where you're trying to meditate, but there's someone in the room playing a didgeridoo

Islander E Ah

Islander C It's wonderful that you did that

Islander F I had to drive into the city, it was a pain in the tukkus

Islander C No but you're always doing that, you're taking these courses

Islander F Well, I always see you at the library!

Islander C I love to read but I always wish I was taking more *courses*

Islander F Well, in this class I was like—must this didgeridoo player be a very attractive young woman? Who I have nothing against, as an older woman I love younger women: except this one. I found her annoying. There was a weird sexual vibe in the air, as if the teacher, who was a man with long hair and very bad body odor, he wove this feather into his hair

Islander E Oh I think I know him he didn't pay me for some / contracting work

Islander F Yes he actually lives on the island part time but anyway, as part of the class we were all supposed to learn whale sounds together, so I actually do speak whale.

Islander C Wow

Islander F Pidgin whale, not well, I mean

Islander C Not *whale*

Islander B Ha!

Islander A Well, what can you say?

Islander F I can say um . . . well, there's not really meanings, well, there's sort of meanings but okay, well, they think this is a sort of greeting . . . . . *blaasssssshhhaaaaaaaorrr* . . . . . or no it's more like . . . *maaaassssorrrrrooooooooo* . . . ah, well, I was never

Islander G So I guess we won't be able to ask them for consent

Islander D *returns with a latte.*

Islander B You're back!

Islander D Yes **Islander A** Hi **Islander F** Welcome back

Islander H What did your friend say, the oceanographer, from Germany

Islander D Well, he wasn't much help.

*

Islander B That at no point in my life have I had certainty at every moment I've been plagued by doubt, at every moment I feel as if others are so clear on what they want to do, as if they're so clear and I am always like what the hell should I do I mean is this clear to other people about what the hell to do about *anything*

*

Islander D Well, what else can we do?

Islander A I don't know, we could call the city

Islander B I called the city they said they're not sure

Islander C The um the EPA

Islander A The EPA has their hands full, there's a lot going on

Islander D Apparently lots of different kinds of animals are going missing

Islander G Is that true?

Islander F That's what they're saying on the news

Islander G Well, that's upsetting

Islander A The people at the university

Islander B They say they're trying to find them but they don't know

Islander C We could go out there in a little / boat

Mayor Annie What, and look for them? How would that help. That's the worst idea ever. That's truly the worst idea ever. I mean that is the worst idea ever. I mean talk about bad ideas: that's one of the worst ideas I've ever heard. I mean that is truly bad. That's a bad one. That's a terrible fucking idea. Go out there in a little boat? *Spit.* I'm sorry. I've lost control of this meeting.

*

John Stay here at my place if you want, Gary.

Gary Why would I need to stay here

John Well, I'm not saying you / need to

Gary I don't need to

John I'm not saying you *need* to I'm saying you *can* stay here, if you want. If you want.

Gary But why would I want that?

John Gary. I have a suspicion that um . . . that you are in fact homeless. That there is nowhere for you to live. I have that suspicion. And you seem like someone who is . . . very proud, and I respect that, believe me, that internal sense of dignity.

Gary People don't need houses. They don't need houses. There's actually lots of ways to live. I'm not homeless. I do good for myself. I've found my own ways.

John Okay, well. Sure. Of course.

Gary Let me know if you have any more wood that needs to be cut.

John Yes I will. I think it will be a little while, because you chopped all of the existing wood. But I certainly / will.

Gary Well, let me know when you *do* have more wood, I'm not saying there's more now

John No no I know

Gary I'm saying let me know when you / do

John Yes. When I do. I will find you.

*

Islander D Hey

Islander A Yeah

Islander D I know this is crazy but I was thinking about taking my boat out tomorrow, just seeing if we can spot any whales—do you want to um . . . do you want to come?

Islander A Oh

Islander D Could be fun, bring a couple beers?

Islander A Yeah, no, that could be um. . . . sure. That could be good. That could be fine.

Islander D Tomorrow afternoon?

Islander A Yeah sure. Yeah no sure, no. No that sounds fun.

Islander D Oh look

Islander A What

Islander D Bioluminescence

The sea is alive with bioluminescence.

*

Leslie Hey haha thought I'd reach out over email just to say that I was thinking of calling but it did seem too late so I'm emailing. I'm just looking out the window at my horses. They're restless tonight they can't sleep I don't know why. I've tried everything I sang to them I brushed their manes I gave them more food I inspected them for fleas I told them how beautiful they are I stood right in their sightline so they'd know I was there . . . but they're so restless *something* is off. Anyway it's been tough to connect with you recently, I mean it's been hard to track you down . . . it *was* wonderful to get that photo of you, you're very handsome, the beard *does* suit you. And I hope that you liked my photo . . . I sent it a couple days or so ago so you should have gotten it . . . and I haven't heard from you since then . . . I've lost a little weight since that photo I think I said, but . . . it's been such a joy to communicate over letters mostly, and some phone calls, and some emails like this one haha . . . anyway I hope you're okay my darling, can I call you that? . . . and I look forward to hearing from you soon . . .

*

Ella (*asleep*) No . . . God . . . please . . . that you . . . off me get . . . GOD—off ME get OFF ME get off you OFF me OFF. . . . . . .AHHH

Ali Mom

Ella They let you on the boat thank God, my sweet

Ali Mom, no it's—you're dreaming

Ella I'm what?

Ali You woke me up, you were calling out

Ella Oh I'm

Ali You're

Ella Sorry I'm confused—bad dream I'm

Ali Do you want some water?

Ella Yes please water

Ali *brings a glass of water.*

Ella *gulps it down—so thirsty.*

Ali Are you in pain?

Ella No worse than usual

Ali Do you want me to come sit in here?
Or I could rub your feet

Ella No that's not necessary go back to sleep I'm sorry to / wake you

*

Ali Fucking waking me up screaming. She is so scared. I mean I would be scared, I get it but she won't *say* it she changes the subject she's fine talking about the past or about *my* life but when it comes to what is actually happening now she—yeah. Won't.
For it to be over.
Horrible to say, but yeah.
I just—miss my life.
I want to get back to my life.
I miss you, actually.
I've been missing you a lot.
It's nice to hear your voice.

Hey I know I said I wasn't coming back to New York after this was over but um—
what if I came back?
Oh.
No, that makes sense. I get that.

*

Ella I was born in Seattle in 1955.
Went to high school at Roosevelt, studied Art History at [the] U-Dub.
Spent a summer in Italy—incredible.
Had my beautiful daughter, sweet daughter.
Moved to the island.
Worked for thirty years as an administrator at the high school, in various roles.
Grew a garden.
Voted in every election.
Threw great birthday parties for my friends.
I was diagnosed with cancer two years ago in June.

*

Alexander Mom

Mo Yes, Alexander

Alexander Can't sleep

Mo Why, honey?

Alexander I'm worried about the dance recital I'm worried I'm not going to be good

Mo Oh, honey

Alexander I've been practicing a lot

Mo Yes you have and it's been wonderful to see

Alexander I'm worried that even though I've been practicing a lot, and really trying to do my best, that I'm not good. I'm worried I'm still not good. Am I good? Mom? Do you think I'm good?

Mo Of course you are

Alexander But not in general, I mean at dancing. I mean am I a good *dancer*?

Mo . . .

*

Gary Follow me

He moves across the stage, trailed by a few, then tens, then hundreds of small dark forms.

Follow me, dogs

Come with me, dogs, come with me
Come with me, you dogs, you wolves . . .
Deeper and deeper . . .
Deeper and deeper and deeper . . .
Deeper and deeper and deeper . . . into the woods . . .

Scene Five

Ella Day I had chosen to die

Ali Mom

Ella Hi

Ali What are you doing?

Ella Just woke up

Ali Do you want some water or anything?

Ella Water here

Ali How are you feeling?

Ella Pain

Ali I'm sorry, I don't know what to do.

Ella [*Come*] —here.

She does.

Ella Nurse called?

Ali She said the ferries are all messed up.

Ella Going to be late?

Ali Probly

They sit there for a moment.

Ali*'s hand is on her mother.*

*

Leslie I've booked a trip to Spain. Yes, well, it was always something I wanted to do. I'm going hiking! I've booked a hotel for each night. I'm going to go in summer, so the grass will be golden, I'll stop for a beer somewhere in the late afternoon, I'll sip my golden beer looking out at the golden grass, with the wind blowing . . . I'm very excited. I'm going by myself.

*

John Santiago. It's me, John. I know it's been years since we but I was thinking that I just have to tell you that I before it's too late that I

*

Mary Hi

Chris Hi

Mary I told you not to come by

Chris I know
You guys eating dinner?

Mary Yeah

Chris What are you having

Mary Macaroni and cheese

Chris My favorite

Mary . . . .
. . .
Do you want to come in?

Chris Yeah

*

Mayor Annie Speech from Mayor Annie.
Well, it looks like our effort to find the whales has failed as of now. But still I think they might come back or some of them might come back if we keep our eyes open. Maybe on certain issues, there's not a great deal that can be done, except to keep your eyes open and be aware.
And I know this isn't an elected position. I know it's not a godamn elected position like President, or Governor, or even member of the US House of Representatives
Or Local State Representative or State Senator
Or even county council member
Or even the water board, for chrissake, which I know is an actual elected position.
But my God: I want the record to show. At my end of first year as Mayor (symbolic Mayor), that although I was elected on a platform of the Humane Society because people like dogs, that I have nonetheless governed with an expanded mandate.
To do good *everywhere* that I could. And that I have thought about personally and tried to get everyone else to think about on a kind of group level about how we can thrive as a community, taking care of ourselves, but also of each other, and also of the natural world around us, which in turn takes care of us, giving us clean air and water, and food and stuff like that. We are not separate from the earth. We are of the earth. I mean I don't understand what's so goddamn hard about that to understand.
So I don't know who will be next symbolic Mayor
But I hope for God's sake that you don't stick to the issue on which you were elected
But that you view your mandate as expansive
As I did mine.

So I guess I'm signing off now. And giving up my powers, or lack of powers.
So in the words of one of the greats:
Good night
And goddamnit:
Good luck.
Sincerely,
Former Mayor Annie

*

Islander D Hey

Islander A Hey

Islander D You wanna take that boat out?

Islander A Yeah sure

Islander D It's a beautiful day

Islander C They take the boat out.

*

John Hello

Ali You came

John Yes

Ali She's sleeping, I'm sorry. It's hard for her to stay awake right now because of the medication. And now we're just waiting for the nurse, there's something fucked up with the ferries

John That's alright, we should let her sleep. Is she in pain?

Ali Last night she was but I gave her a lot of the pain medication and this morning when she woke up she seemed pretty um blissed out or calm, I guess, relatively, although she did tell me she was in pain.

John I hate knowing she's had to experience all this pain.

Ali You brought flowers!

John I hope that's okay. I know she didn't want me to come and I wanted to honor that but I thought. I looked out at my garden and I thought: I have all these flowers. I was supposed to take them into the city to sell at the Pike Place Market. I thought it would be good to have something to do today, to keep on working. But then I thought goddamnit. My friend is dying today. My lifelong friend. And goddamnit I don't want to go to work. I don't want to not come just because she asked me not to come. Fuck her wishes. I matter too. And what I want is to cut all these flowers from my garden and bring them to my friend. To my lifelong, my dear, maybe my dearest—yes—friend. And that's what I'm going to do. So fuck the market, and fuck her. Anyway I have all these flowers now. Will you help me bring them in?

Ali That's very kind of you

John I'm warning you –there's really a lot. I started cutting and I couldn't stop.

They bring in bucket after bucket of flowers.

*

Alexander Mom mom watch me
I'm nervous for the performance today, for everyone to be watching me
Can I do the dance for you, before we leave?

Mo We don't have much time, Alexander, we have to get there

Alexander Just once, please
To practice

Mo Alright, just once
And then we have to go
But don't wear yourself out
Save it for the real thing!

*

Ella I'm awake
I can see the flowers
I can see outside the room
The house . . . the garden . . . the road . . .
Luckily I can see through my eyelids now
Luckily I'm hearing through my fingers now
Luckily my head is opening towards the sky

*

A whale enters.

No special costume.

Whale I am a whale, and I have always been a whale.
There's nothing cute about it.
If I could somehow show you how big the ocean is, and how great it is to be inside it.
Everyone says it's getting polluted, and that's true. The trash is okay, we don't notice that too much. Even the temperature is okay, the changing temperature. It's making some of us sick, and affecting our food supply, and that's not so good. But we can deal with that. Death is no problem, if all of us die, that's no problem, not for us.
What scares me is the loss of silence. There are more noises now, and I hate that.
Weird booms in the distance, engines, you have no idea how far the sound of engines travels, when you're a whale underwater.
But still there are stretches of the ocean where you can just swim and swim, and hear nothing except distant fish and
Faint rustles and movements on the surface of the ocean floor

Very far below you.
My group of whales and me used to go every year to the water outside the island.
It was a tradition.
But this year we split up, we were like, let's just do our own thing.
Mothers split from their kids, friends were like take care, maybe we'll cross paths again down the road.
I wandered around for a while.
And then I decided to swim back up towards the island. I'm nostalgic, I guess, and I missed it.
There's something about the water there that's so beautiful.
I swam for a long time to get there.
And finally I saw it
And as I got closer and closer
I remembered what it was like to be a kid there, when my mom was around me, and my brothers
What fun it was to be a whale
My swimming got faster and faster
And all of a sudden
I was overcome with a feeling I could not describe
And it was so overwhelming
That I couldn't help it

Islanders D *and* **A** *see something in the water*

Islander D Hey, look!

Whale I leapt into the air.

Somehow, the whale leaps from the water, his massive body turning in the air, just as:

Something happens to **Ella**, *just as:*

Alexander *does the last few moves of his dance.*

He's dancing his heart out—he's really been practicing.

He finishes, breathing hard.

Alexander Mom
How was it?

Mo Oh, honey

Alexander How *was* it

Mo Oh, honey
My darling
It was

It's more than she can express.

End of play.